Dimensions of Time and Location

*A Different View on the (im)Possibilities of Travelling
through Time across Locations and Space*

Time

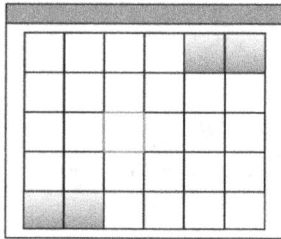

Date

Location

Clemens Willemsen PhD

ISBN: 979-88-6053-121-5
Printed and published by: Kindle Direct Publishing
Copyright: Clemens Willemsen 2024

Table of Contents

List of figures

List of tables

1. Foreword

Your life and the world we live in might seem to be totally planned or everything happens by what seems coincidence. Those are two extreme opposite philosophies. Is it a coincidence in what year and what part of the universe you were born and how your life is shaped? Does something like self-determination exits or are we subject to whatever comes on our path? This is not a book on religion or philosophy but on a different view on the dimensions of time and location.

People have been intrigued by the dimensions of time and location[1]. You were born on a certain date, at a certain time and at a certain location. All these elements are fixed or seem to be fixed and they are also used for identification purposes. That comes with the 'dimensions' of lineage (or descent), culture and other aspects that influence or determine your life certainly when you are an infant. This book is about the (im) possibilities of traveling in time and or location. Not all aspects will be explained or described here as there are already books, movies and numerous texts on the internet about this subject for which I will make a reference list in attachment 1. It is not very useful to quote large texts from the internet in this book as you can follow the links. Another reason to limit the scope is that I am not a physicist, biologist, cosmologist or a space expert so themes like black and worm hole are not described in detail. You can find information on these separate subjects in different sources but my aim with this book is to relate these various subjects and conclude some principles for time and location travel.

Travelling in time and/or location is not completely separated as can be shown in the next figure. Travel in a cocoon or pod in a space ship for example is a combination of both dimensions as you will travel in space[2] for a long time.

---

[1] The pictures on the cover are free clip arts from MS Word.
[2] Locations can be seen as points in space and space can be regarded as the area of locations.

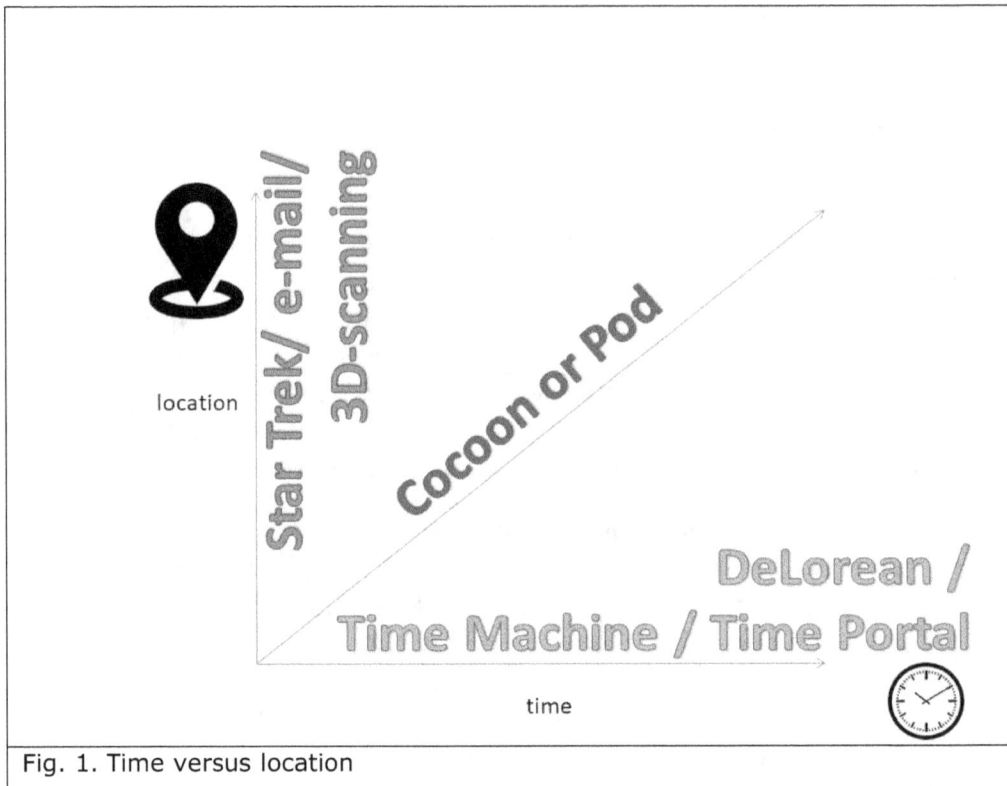

Fig. 1. Time versus location

I will end this book with the example of traveling to Mars where I will go into detail about the consequences seen from the dimensions of time and location.

A different subject but somehow related: Unseen worlds

In the past aristocrats and other wealthy families had their business taken care of in an unseen way. I refer to the servant who would do the cleaning in the house and preparing or serving food. They were not supposed to be seen by the family so there were separate hidden staircases and corridors with separate doors to the main rooms. You might remember the British drama tv series of 'Upstairs, Downstairs' in the 70's or the sitcom of 'You rang, MLord?' in the 90's. The family would use the regular broad staircases and doors and the servants would not be seen doing their work. Servants would go up and down on these separate, narrow and steep staircases from the cellars with the kitchen and laundry room to their sleeping facilities in the attic. You might recognize the entrance to such a staircase by what does not seem to be a door and can be wallpapered so it is not obvious. Two different worlds I would say! In the dimensions of time and location you can say that the servants had their own locations (bedrooms, kitchen and staircases) all the time and were allowed in the family locations (living rooms, family bedrooms) sometimes only to do their job.

I can draw a parallel to an unseen world today as well. You do not have to go to a store to pick out an item, pay the cashier and take it home or have it delivered at home but you order it on line. You can follow the whole process from the warehouse by the truck to your house and it might be delivered at a special outdoor (refrigerated) box of your house at a DHL box, at the grocery store or a pharmacies location. I consider this as different routes for information on a product or service and routes for the physical product itself. Modern technology has made it possible to separate the two routes where they were one in the past. I will explain this more in detail further on at the topic of e-mail.

A little deviation; The use of Wikipedia for scientific research and publications

Writing my dissertation I tried to stay away from using Wikipedia as a source and referring to the webpages. In that time period from 2011 till 2019 I also borrowed printed books from (academic) libraries like the Department of Justice and Security, Tilburg University and the Police Academy. Also books got more and more available as an electronic book or portable data format (pdf) which was more convenient for searching and storing your own digital library. Referring to a book or an article in a magazine was and is considered to have more scientific value as these books and articles have to go through a process and you can easily find out who the author was or the authors were.[3] In my more recent books I have not followed this guideline so strict for a number of reasons. Sometimes you cannot find a book at a bookstore or library and then Wikipedia is the only source. Wikipedia shows the revision history of their webpages so you can see who has changed what content. And as I know from my own experience, you can publish a book online by yourself without an editor involved so hard cover books are not always more 'reliable'. The best way is to use various sources to find the truth.

---

[3] You can find out more on the reasons not to use Wikipedia on
  https://en.wikipedia.org/wiki/Wikipedia:Academic_use.

2. The Dimension of Time

2.1.   Introduction

You can travel In time that is forwards just by waiting till the next day, month or year appears but mostly we regard time travelling as fast forward or fast backwards in time like with a 'time machine' or a cocoon or pod in a spaceship like in the movie 'The Passengers' that preserves your body and awakes you at a later moment in life and a changed environment even at the same location. It would be weird to wake up as you have been dormant and people as well as the world around you have changed in the meanwhile. Would you be able to adapt after a while? Think about Japanese soldiers on a remote island in the Second World War who were discovered years after the war had ended. There was also the movie of 'Back to the future' which used a DeLorean DMC-12 car as a time machine. If you move backwards in time you might pass the day of your birth and those of you parents. What if you do something and lives of people change? Can you rewrite history? Then there is a time loop like in the movie 'Ground hog day' when a period of time is repeating itself for the main character only and he cannot get out of the loop.

In a way we can all travel back in time as we can hear audio recordings from the past and look at movies. Just think about the coronation of King George in 1937 that we still watch today. Especially in 3D or virtual reality it appears like you are part of it and can relive the old times.

2.2.   The Concept of Time

Time is an interesting object that we take for granted but it is good to see what it is really all about. When we talk about time we do not just mean the time in hours and minutes but the time including the date as in a timestamp. We use the concept of second, minute, hour, day, 24 hours ('nychthemeron'), week, month and year but where does this come from? I guess that early mankind knew the element of a day as two occurring aspects when the sun was at its highest point and did not need a calendar. Of course now we know that the highest point of the sun depends on your location and changes during the year as the earth moves at an angle around the sun in a circle. To man it might appear in those days that the sun turned around the earth. In Dutch like in some other languages we have a common word for this and we speak about an 'etmaal' or one day and one night combined meaning 24 hours in English but we have not discussed the concept of hours yet. The proper word in English is 'nychthemeron' as a loanword from Greek and appears in the New Testament but it is not used commonly. A month was determined by the position of the moon which also causes high and low tide during the nychthemeron. A year must have been known in early times as well considering that the change of seasons with temperature, sunshine and growth repeats itself in about 365 days. Man was a hunter and a gatherer where plants as well as animals have a pattern that is repeated during the year as nature changes. I think that a minute, hour and week did not make much sense to a prehistoric man as there were no methods to measure 'time' in the old days. How did it occur that throughout the world we determine an hour as a period of 60 minutes with each 60 seconds and a week consists of seven days? A week consists of 7 x 24 x 60 x 60 seconds or 604,800 seconds. Therefore you could think about another timescale for the week which is more decimal like:

days	hours	minutes	seconds	total seconds in a week
10	20	50	60	600,000
7	24	60	60	604,800
6	20	50	100	600,000
6	20	100	50	600,000
Tab. 1 Timescale for the week				

Especially when we use 100 seconds in a minute or a 100 minutes in an hour, we could more easily speak of ½ hour or 0.5 hour as 50 minutes. That would make calculations easier.

You can compare this decimal timescale in a way to the scale of the old British pound or pound sterling consisting of 20 shillings where each shilling was 12 pence making a pound worth 240 pennies. After the decimalization in 1971 the British pound consists of 100 pennies.

Ancient Greece had ten days in a week and the Etruscans had eight days. Nowadays the seven days in the week have their own names after names of gods or planets visible to the eye[4]. It is not clear what the 8th, 9th and 10th day were named in Ancient Greece.

Day	Monday	Tuesday	Wednesday	Thursday	Friday	Saturday	Sunday
Planet	Moon	Mars	Mercury	Jupiter	Venus	Saturn	Sun
God	Mani	Tiu	Odin	Thor	Frigg	Saturn	Sol

What is considered the first day of the week differs like:

method	first day	used in
ISO standard 8601	Monday	EU (excl. Portugal) and most of other European countries, most of Asia and Oceania
Broadcast calendar	Monday	Broadcast USA
Western traditional	Sunday	Canada, United States, Iceland, Portugal, Japan, Taiwan, Thailand, Hong Kong, Macau, Israel, Egypt, South Africa, the Philippines, and most of Latin America
Middle Eastern	Saturday	Much of the Middle East
Tab. 2 First day of the week		

The ancient Egyptian calendar – a civil calendar – was a solar calendar with a 365-day year. The year consisted of three seasons of 120 days each, plus an intercalary month of five epagomenic days treated as outside of the year. Each season was divided into four months of 30 days. These twelve months were initially numbered within each season but came to also be known by the names of their principal festivals. Each month was divided into three 10-day periods known as decades. A decade is nowadays a period of ten years. This civil calendar ran concurrently with an Egyptian lunar calendar which was used for some religious rituals and festivals. The Egyptians appear to have used a purely lunar calendar prior to the establishment of the solar civil calendar. The lunar calendar divided the month into four weeks, reflecting each quarter of the lunar phases. The days of the lunar month — known to the Egyptians as a "temple month" were individually named and celebrated as stages in the life of the moon god. Each day of the lunar month had a unique name or better Egyptian hieroglyph different form the unique names of the days in the week we use know and who repeat itself within a week and not within in month[5].

---

[4] https://en.wikipedia.org/wiki/Week

Half of these days have a specific meaning mostly regarding to the phases of the moon like:

1	2	3	4	5	6	7	12
New moon	Crescent moon	(first) Arrival	Going forth of Sm, a priest	Offerings upon the altar	Sixth	First quarter day	Partial second quarter day
15	16	17	18	23	26	28	30
Full moon	Second arrival	Second quarter day	Moon	Partial third quarter day	Going forth	Jubilee of Nut	Going forth of Min

For much of Egyptian history, the months were not referred to by individual names, but they were numbered within the three seasons after the water situation in the river Nile.

I	II	III	IV	V	VI	VII
1st month of flood	2nd month of flood	3rd month of flood	4th month of flood	1st month of growth	2nd month of growth	3rd month of growth
VIII	IX	X	XI	XII	intercalary	
4th month of growth	1st month of low water	2nd month of low water	3rd month of low water	4th month of low water		

We need to know this concept of time in order to look more at the phenomena of time travel. A universal denotation of time (with date) is needed like: year-month-day-hour: seconds. One for the start and one for the destination is needed. The year[6] has to be specified with BC (or BCE) or AC (or CE)[7] in the Gregorian calendar that we use in the western world today. 'Calendar' comes from the word for the first day of the month according to the Roman calendar. There are more calendars in use like the Islamic that started when the prophet Muhammed moved to Medina and now has the year 1445 or the Roman calendar that was based on the reforms introduced by Numa Pompilius and started in 713 BC.[8]

---

[5] https://en.wikipedia.org/wiki/Egyptian_calendar
[6] A light year is in fact not a year but the distance the light travels within a year so a measure for travel and location.
[7] Before Christ or Before Common Era and Anno Domini or Common Era.

Time zone

ISO 8601 is the international standard on date and time related data. There are a couple of representations where for example "2007-04-05T14:30Z" is commonly used. T stands for Time and Z for zone designator. A time zone is actually a combination of the dimensions of time and location (or space) on which we will come back further in this story. The Z also means that the local time is in the zone of UTC or Coordinated Universal Time as the successor of Greenwich Mean Time (GMT). A local time is an offset from UTC. "2007-04-05T14:30Z" (Greenwich) is the same as "2007-04-05T15:30+01" (Amsterdam) according to my calculation not regarding summer/wintertime. Another representation of the day is by its number in the year like "1995035" for "19950204" or February 4, 1995. If we look at the earth from the universe then the UTC would be the most likely zone to consider as the 'time at earth'.

A principle is: **you must use a standard notation for time and day including the time zone**.

One principle on time travel could be that you can observe the past or the future but you cannot change it. Let's assume you travel to the future and see on what lottery number a price has been drawn. Then you could go back to the present and buy that lottery ticket and become a rich man. But the next time you travel to the future you would be wealthier than the first time you traveled to the future. Or does that mean that many futures are possible and you can influence them? Another event would be that you travel to the past and prevent your father and mother to meet. Than you would not be born which is in contradiction to the present situation. So we can derive a principle as: **you cannot change or create an event in the past or in the future.** The question remains if you could get knowledge on the future like the winning lottery ticket and remember that knowledge if you return to the present but not do anything with this knowledge? This would lead to the principle of: **you cannot remember what happened in the future or the past at the moment you were there**. It would be even less than a dream that you can remember of fragments of that dream. We could change this principle into: **you cannot remember details of the future or the past at the moment you were there**.

---

[8] See https://wonderopolis.org/wonder/what-do-bc-and-ad-stand-for

2.3.   The Means of Time Travel

Time travel contains usually a person or a small group of persons who travel to a specific moment in time. That makes sense as we talk about transportation which is limited by capacity. What if the whole world returns to a previous time? Then we would not have the issue of a surrounding that is strange to us or the other people as we are all in the same situation. In a way this happens when we switch from summer time (standard time or daylight saving time) to winter time at 3 o'clock in the morning on a certain date that differs from continent to continent. We relive that one hour that already has passed. You might state that this is not really time travel as time actually continues but the registration of time by man is interrupted; you have not gained an hour. You can also compare this to travelling west by plane and arriving 'earlier' on your destination but as stated before location is part of the timestamp so neither here is a time gain involved.

There are roughly two means of time travel. One way consists of where the body is transported and another way where the body is in i.e. a cocoon, pod, machine or car and the cocoon is transported. The last option seems in a way unrealistic as you could arrive in the past when none of the materials that the vehicle was made or existed or has not been invented yet. Some movies have as a principle that only the body can be transported in time and not any clothes or other objects like in the movie Terminator. Authors can describe time travel but is there any scientific logic behind it?

Time Portal
A time portal is written on in books and used in movies.[9] A great difference with a time machine is that a time portal does not allow you to determine the date and time where you want to travel to. The portal might disappear and not be available anymore. There is hardly any description on how a time portal is supposed to work. You just enter the (circular) portal on one side and if you are lucky appear at a portal on the other side. A time machine is mostly manmade and a portal is more

---

[9] https://en.wikipedia.org/wiki/Time_portal

of a natural phenomenon or according to the story created by an alien race.

Time Machine

'*The time machine'* by H.G. Wells talks about four dimensions, three of which are called planes of Space and the fourth dimension is time. These planes or directions of a body (p. 4) have length, breadth and thickness each at right angles to each other. Nowadays we would talk about length (or depth), width and height of a body measured for a central point (0, 0, 0). We cannot visualize the fourth dimension which also is at a right angle with the other three dimensions according to the time traveller in the story. These three dimensions remind me of mathematics at high school where we used the X, Y and Z axis with three Cartesian coordinates. More axes or dimensions are possible as mentioned by the time traveller but the fourth dimension is not commonly related to time. The question arises in the book: *And why cannot we move in Time as we move about in the other dimensions of Space?'* (p. 6). We will explore more on this in the part of 'Spacetime'. The time traveller states that he can travel back in time in his thoughts to an earlier moment. There is not much else on explanation of how travel and the time machine would work in the first chapter of the book. If we compare these three dimensions of space to the concept of location I use in this book, then there are some differences:

- H.G.Wells uses the dimension of length in space as the size of the body compared to a corner point at a certain location
- I use the dimension of length in location as the distance from a central point on the earth

Delorean – Back to the Future

"*The control of the time machine is the same in all three films. The operator is seated inside the DeLorean (except the first time, when the remote control is used), and turns on the time circuits by turning a handle near the gear lever, activating a unit containing multiple fourteen- and seven-segment displays that show the destination (red), present (green), and last departed (yellow) dates and times. After entering a target date with the keypad inside the DeLorean, the operator accelerates the car to 88mph (142 km/h), which activates the flux capacitor. As it accelerates, several coils around the body glow blue/white while a burst of light appears in front of it. Surrounded by an electric current similar to a Tesla coil, the whole car vanishes in a flash of white/blue light seconds later, leaving a pair of fiery tire tracks. A digital speedometer is attached to the dashboard so that the operator can accurately gauge the car's speed."[10]*

There is no explanation how the time machine works.

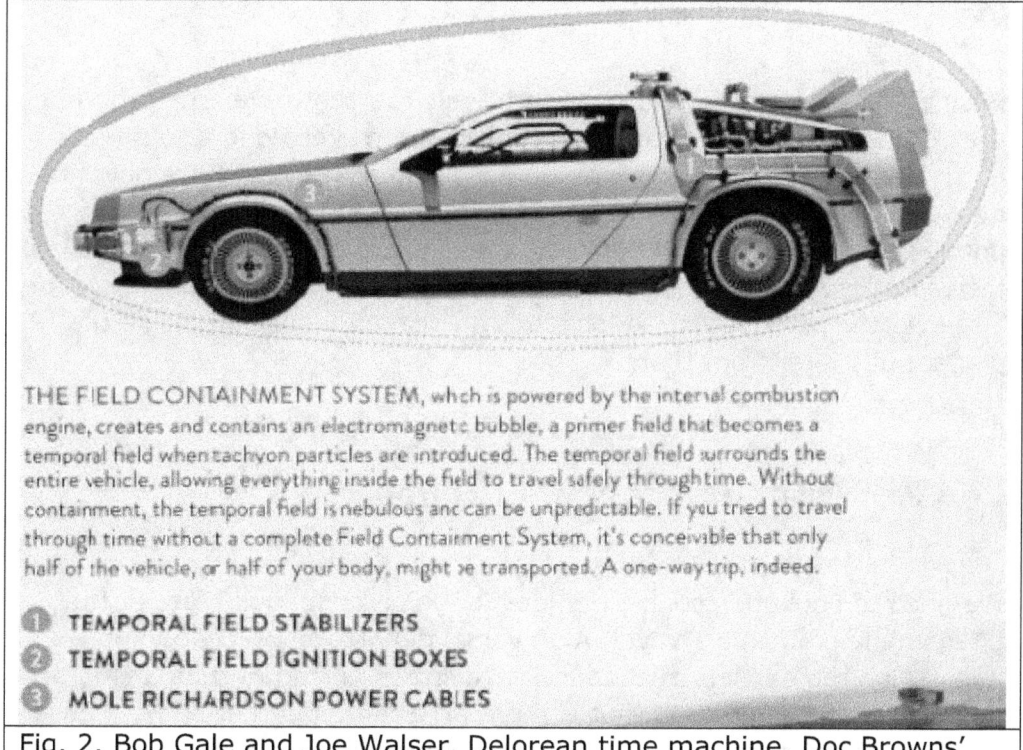

THE FIELD CONTAINMENT SYSTEM, which is powered by the internal combustion engine, creates and contains an electromagnetic bubble, a primer field that becomes a temporal field when tachyon particles are introduced. The temporal field surrounds the entire vehicle, allowing everything inside the field to travel safely through time. Without containment, the temporal field is nebulous and can be unpredictable. If you tried to travel through time without a complete Field Containment System, it's conceivable that only half of the vehicle, or half of your body, might be transported. A one-way trip, indeed.

1 TEMPORAL FIELD STABILIZERS
2 TEMPORAL FIELD IGNITION BOXES
3 MOLE RICHARDSON POWER CABLES

Fig. 2. Bob Gale and Joe Walser, Delorean time machine, Doc Browns' Owners workshop manual, p. 51

---

[10] https://en.wikipedia.org/wiki/DeLorean_time_machine

3. The Dimension of Location

3.1.   Introduction

Besides time travel there is location travel. You can travel in location that is just by walking or taking a vehicle and travel along a route on earth but mostly we regard location travelling as moving extremely fast between two distant locations like at "Beam me up Scotty" in the popular Star Trek series. I prefer to speak of travel between location instead of space travel as the first term is broader and includes travel on and over the planet earth. A location can also be a position inside the planet but I refer from that.

3.2.   The Concept of Location

We need to know this concept of location in order to look more at the phenomena of location travel. A universal denotation of location or address on the earth is needed like: X degrees north (latitude), Y degrees west (longitude). That is more precise than street name, house number, postal code, town (,state,) and country. *'For example, the Empire State Building is located at 40.7 degrees north (latitude), 74 degrees west (longitude). It sits at the intersection of 33rd Street and Fifth Avenue in New York City, New York, United States.'*[11] An address is more alterable than the combination of latitude/longitude as the name of a street might change or disappear. For instance, my birthplace became part of a different municipality after years so that is not a constant factor especially if you want to retrieve your birth certificate. If a street is named after a famous person then the name might be changed if that person is no longer in favor. Longitude and latitude might not be known if we travel far in the past and the landmasses on the earth move so this is neither an absolute measure for the location.

---

[11] https://education.nationalgeographic.org/resource/location/

One location for the start and one for the destination is needed for travel.

The above denotation is limited to our planet Earth. Travels have been made to the moon by men and to Mars by instruments only so this would be something to consider at a later moment to get a complete location address like solar system, planet, latitude and longitude. Addressing the Earth as the third planet from the sun at a distance of about 150 million kilometers might be precise enough for this book.

3.3.   *The State of Body and Mind*

Traveling by location asleep in a cocoon in a space ship might have a great impact on your body and mind and can be compared in a certain way to hibernation. Animals hibernate like frogs. Bears do not really hibernate but sleep. A patient might be put in an induced coma or be in a coma. A coma might have a negative influence on the aging of your mind and body. The ability of the brain during a coma to regulate the metabolic process is compromised and might lead to imbalances in hormone production, immune function, and cellular repair mechanisms[12]. The question is if a coma would shorten your life span i.e. if you travel in a space ship in a cocoon for 100 years at the age of X would you still be X years old or X plus a part of those 100 years? The older you get, the more functions are lost from the cells in your body[13]. Travelling for ten or twenty years in a coma would certainly have a negative impact on your mind and body I assume from this. The eldest person ever became 122 years so there is a limit to how old you can get. How would you measure the 'time' that has passed? Then it would defeat partly the reason to travel if it would not elongate your life and you would become much 'older', physically and or mentally. The effect on your mind and body during the period that you're in a cocoon has to be compared to the effect in wear and tear that would occur if you lived your regular life on earth. Is there a method to preserve your body during this travel period? I am not a biologist and therefore I will not go into detail on the

---

[12] https://medicalhealthauthority.com/info/do-you-still-age-in-a-coma.html
[13] https://medlineplus.gov/ency/article/004012.htm

effect of aging on the body cells but it is relevant to know this is a fact and has to be taken into account.

We can compare these states of the (animal[14]) body and the mind based on common knowledge.

state of mind	regards	period	purpose	characteristics	activity
sleep	person in a cocoon in a space ship	months or years	rest the body and mind, restore, travel long distances to other planets	reduced body heat, reduced breathing, active brain, dreams	REM, little to much movement, less muscle activities
estivation[15]	reptile and insect	hot, dry season	survive in a harsh environment in summer with high temperatures	reduced body heat, reduced breathing, light state of dormancy	none
(deep) hibernation	reptile or primate	months	survive in a cold season (winter) and absence of vegetation by conserving energy	reduced body heat, reduced breathing	none
(light) hibernation or torpor[16]	mammal like a (black) bear	months or just part of the day or night	survive in a cold season (winter) and absence of vegetation by conserving energy	reduced body heat 3-5 degrees Celcius, low metabolic rate, reduced breathing, bears are able to recycle their proteins and urine	give birth for the females

---

[14] Plants can also have a dormant state but that is not comparable to animals.
[15] https://en.wikipedia.org/wiki/Aestivation
[16] Torpor is seen as the inactive time during a longer period of hibernation.

(medically) induced coma	person	days or weeks	let the body heal before operating by reducing the metabolic rate of brain tissue[17]	high risks losing respiratory drive, bed sores as well as infection from catheters. Requires mechanical ventilation	none
coma	person	months, years or decades	none (except for medically induced), result of accident, stroke or trauma	a deep state of unconsciousness, irregular breathing	unable to consciously feel, speak or move. Some reflexes

Tab. 3 States of the body and mind

BBC news reported in December 21st, 2006 on a Japanese man who survived for 24 days in cold weather without food and water after mountain climbing. *"He had almost no pulse, his organs had shut down and his body temperature dropped to 22C (71F) when he was found. He fell into a hypothermic state at a very early stage, which is similar to hibernation, according to Dr Shinichi Sato, who treated Mr Uchikoshi. He had to be treated for severe hypothermia, multiple organ failure and blood loss."*[18]

Reasons why it would be useful for humans to be in hibernation are[19]:

- *"saving the lives of seriously ill or injured people by temporarily putting them in a state of hibernation until treatment can be given.*
- *for space travel, such as for missions to Mars".*

Would it be possible to use any of these methods for a deep sleep in order to travel in time and location? In order for animals to go into hibernation, they must eat a lot before, create fat and so build up spare energy to be used during the time of hibernation. For a human being it is not unthinkable that the energy is supplied by a feeding probe. An induced coma has many risks and requires detailed attention by experts. I'm not mentioning the option of cryonics like in the movie 'Alien' where

---

[17] https://en.wikipedia.org/wiki/Induced_coma
[18] http://news.bbc.co.uk/2/hi/asia-pacific/6197339.stm
[19] https://en.wikipedia.org/wiki/Hibernation

the body is kept at a very low temperature of – 200 °C as this causes even more damage to the mind and body than hibernation.

The European Space Agency (ESA) seriously thinks about hibernation and torpor as an efficient and plausible way to reach the planet of Mars.[20] It would safe on costs, reduce the size of the spacecraft by a third and keep the astronauts healthy. A 'normal' return flight would need a two years supply of water and food for each astronaut or 30 kilograms a day. And the question would be how you keep the persons entertained so they won't get bored. The metabolic rate could go down to 25% of the normal value. Just like a bear, an astronaut would have to put on a certain amount of fat on the body. There is a however the possibility of a major loss of muscle, bone strength and more risk of heart failure; this means a medical monitoring is needed on the hormone levels during the trip and at the awaking.

A principle is: when travelling in a cocoon for a long period, the person must be monitored for health and be fed (orally) before and or during the trip.

There is also the issue of gravity as we travel from the earth to another planet. Being exposed to less gravity for a longer period has a side effect on your limbs and other parts of your body. Therefore it would make sense to create artificial gravity in a space ship in the shape of a rotating living quarter. The Nautilus-X has been designed in 2011 to do so for a crew of six persons for a duration of one to 24 months[21]. It has only been a concept so far. This would reduce the gravity from earth at 1g to 0.69g where Mars has a gravity of 0.38g[22].

---

[20] Hibernate for a trip to Mars, the bear way, ESA/Science & Exploration/Human and Robotic Exploration.
https://www.esa.int/Science_Exploration/Human_and_Robotic_Exploration/Hibernate_for_a_trip_to_Mars_the_bear_way#:~:text=Bears%20seem%20to%20be%20the,before%20falling%20into%20a%20slumber.
[21] Non-Atmospheric Universal Transport Intended for Lengthy United States Exploration https://en.wikipedia.org/wiki/Nautilus-X
[22] https://space.stackexchange.com/questions/42067/does-lower-gravity-on-mars-make-it-unsafe-and-unhealthy-for-humans

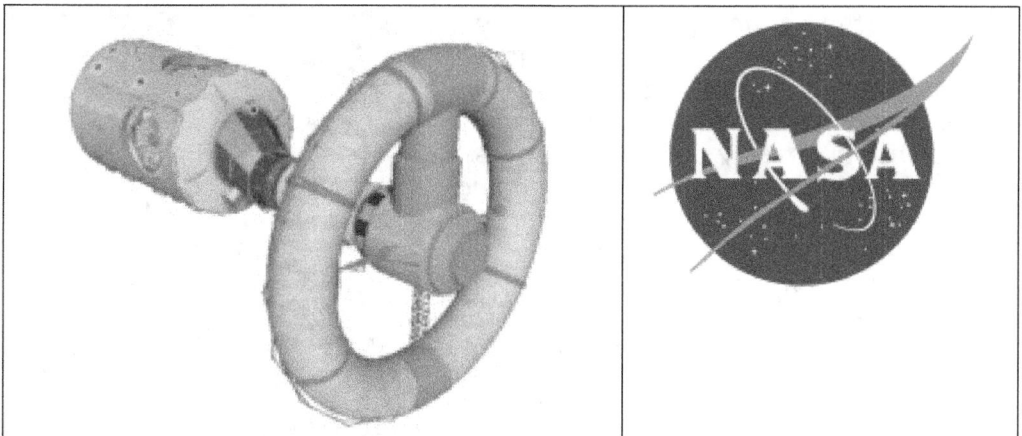

Fig. 3. Nautilus-X https://en.wikipedia.org/wiki/Nautilus-X

3.4. The Means of Location Travel

I mention a few means of travelling to distant locations.

Star Trek - Transporter

If you look at an episode of Star Trek you notice that the command 'beam me up' was given speaking to a communicator or what we would call a mobile phone nowadays. The transport would take place from the spaceship to a planet or the other way around to explore or get away from danger. The space ship had a specific area and the person had to stand on one of six circles like a projection or a hologram of a tube. Mr. Scott was the technician who had to fix the beamer at a malfunction.

A transporter is a fictional teleportation machine used in the Star Trek science fiction franchise. Transporters allow for teleportation by converting a person or object into an energy pattern (a process called "dematerialization"), then sending ("beaming") it to a target location or else returning it to the transporter, where it is reconverted into matter ("rematerialization").On Star Trek: The Original Series, the transporter was portrayed as a platform on which characters stand before being engulfed by a beam of light and transported to their destination. The television series and films do not go into great detail about transporter technology. According to The Original Series (TOS) writers' guide, the effective range of a transporter is 40,000 kilometers. [23]	 *https://upload.wikimedia.org /wikipedia/en/thumb/c/c7/Tr ansporter2.jpg/220px-Transporter2.jpg*
Fig. 4. Star Trek Transporter	

The actual transport sequence is conducted by the auto sequence programs of the transporter controller, under supervision of the operator. The major stages of the transporter operations are[24]:

stages	function
Target scan and coordinate lock	Programming the destination coordinates with automated diagnostic procedures
Energize and dematerialize	Derive a pattern image by molecular imaging scanners of the person where the coils convert the subject into a matter stream.
Pattern buffer Doppler compensation	Compensate for the Doppler shift between the ship and the transport destination. The pattern buffer also acts as a safety device in case of system malfunction, permitting transport to be aborted to another chamber.
Matter stream transmission	One of seventeen emitter pad arrays transmits the matter stream within a beam to the transport destination.
components	*function*
Transport chamber	The protected volume within which the actual materialize /dematerialize cycle occurs.

---

[23] https://en.wikipedia.org/wiki/Transporter_(Star_Trek)

[24] Star trek, The next generation Technical manual. Chapter 9.0 Transporter Systems, p. 102-109 Rick Sternbach and Michael Okuda 1991

Operator's console	The control station permits the Transporter Chief to monitor and control all transporter functions.
Transporter controller	Computer sub processor manages the operation of transporter systems, including auto sequence control.
Primary energizing coils	Create the Annular Confinement Beam (ACB), which creates a spatial matrix within which the materialize/dematerialize process occurs.
Phase transition coils.	These devices accomplish the actual dematerialization /materialization process by partially decoupling the binding energy between subatomic particles.
Molecular imaging scanners	Not relevant.
Pattern buffer	Delays transmission of the matter stream so that Doppler compensators can correct for relative motion between the emitter array and the target.
Biofilter	Scans the incoming matter stream and looks for patterns corresponding to known dangerous bacteriological and viral forms. Upon detection of such patterns, the biofilter excises these particles from the incoming matter stream.
Emitter pad array	Transmits the components of the transporter ACB and matter stream to or from the destination coordinates.
Targeting scanners	Determine transporter coordinates, including bearing, range, and relative velocity to remote transport destinations.

Tab. 4 Stages of the Transporter operations

The normal operating range is approximately 40,000 km, depending on payload mass and relative velocity.

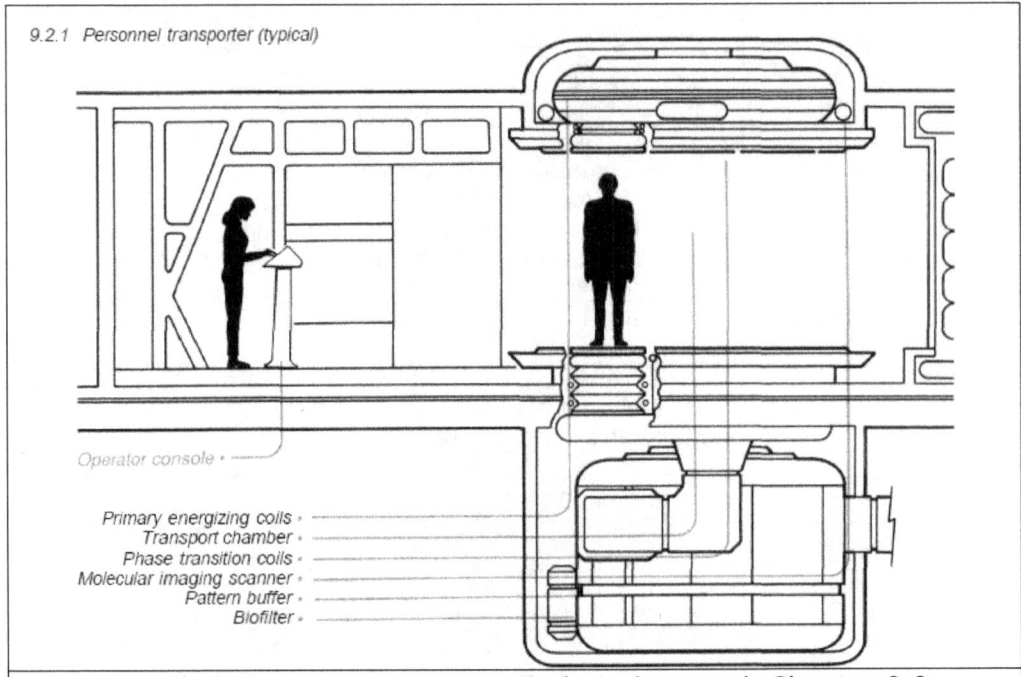

Fig. 5. Star trek The next generation Technical manual. Chapter 9.0 Transporter Systems, p. 103 Rick Sternbach and Michael Okuda 1991

What I do not understand is that your body occupies a certain physical space and if you move to another location, what happens if that another location is already occupied by for example a wall or another person or just air? That seems to be a little illogical. And there is an issue if your body is sent in parts or molecules from point A to point B what if the parts don't all arrive? Another question I have is why you only need one transport on the location you are sent from and not another transporter at the site where you go to?

A principle is: **the location of destination must not be occupied by a subject and also needs a receiving station.**

Cocoon or Pod

ESA (see article before) suggests a pod to be like a soft shell; "*a quiet environment with low lights, low temperature – less than 10 °C – and high humidity. The astronauts would move very little, but would not be restrained, and wear clothing that avoids overheating. Wearable sensors would measure their posture, temperature and heart rate".* This is a little different than the cocoon shown in the figure of the article (see before). I think a soft shell can cause more injuries when moving abruptly than a hard shell and is more based on the nowadays sleeping bags of the astronauts.

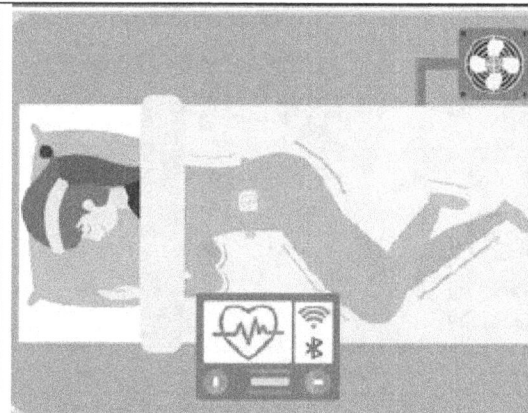

The steps of hibernation would be:
1. enter soft shell, shielded pod
2. administer drug for metabolism
3. initiate torpor
4. wake up
5. rehabilitate

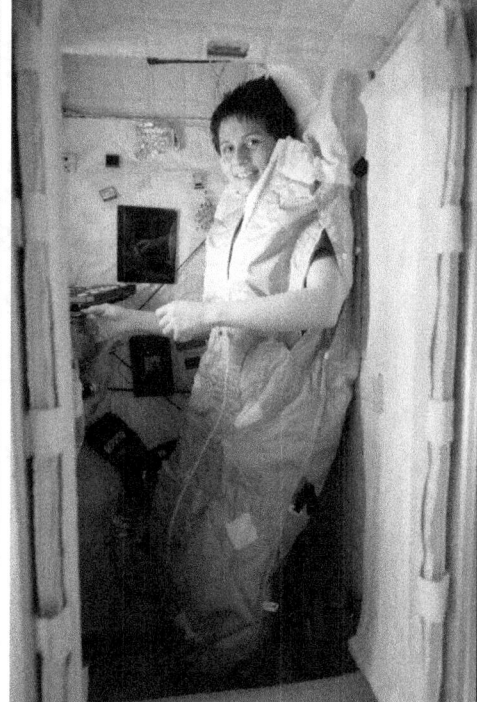

Fig. 6. Hibernate for a trip to Mars, the bear way, ESA/Science & Exploration/Human and Robotic Exploration

Transport like e-mail

We might look at electronic mail messages and how that takes place. It is not quite the same as a message is copied from location A to B in packages and once it is determined that all the packages have been moved, the original message can be deleted and thereby sent from A to B. Location in this way is not the same as location that we talk about in a wider sense as it may even take place on the same computer and thereby on the same location; so not location travel in that sense. We actually talk about a logical location instead of a physical location as meant in this book so far. A logical location can be an ip-adress. An Internet protocol address is a label connected to a computer network and uses a network address which you can compare to a phone number.

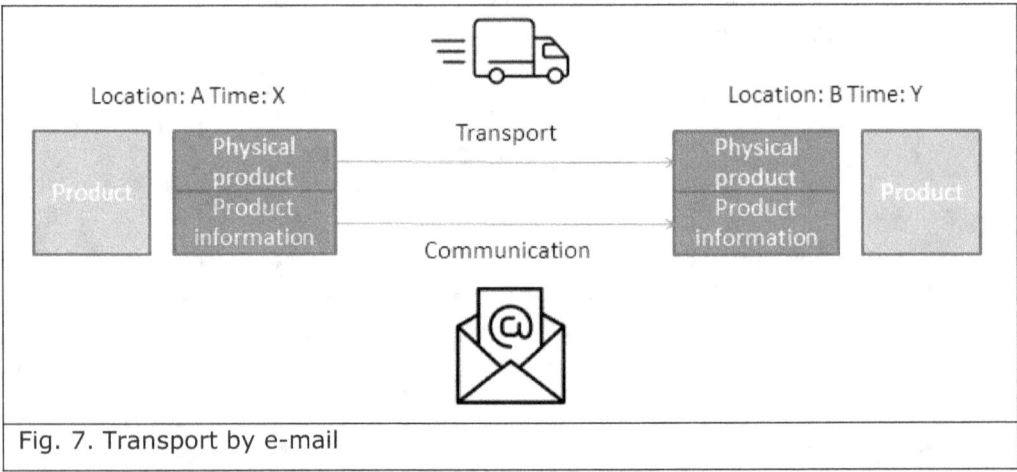

Fig. 7. Transport by e-mail

This is very well explained by Sandip Roy and visualized as follows[25]:

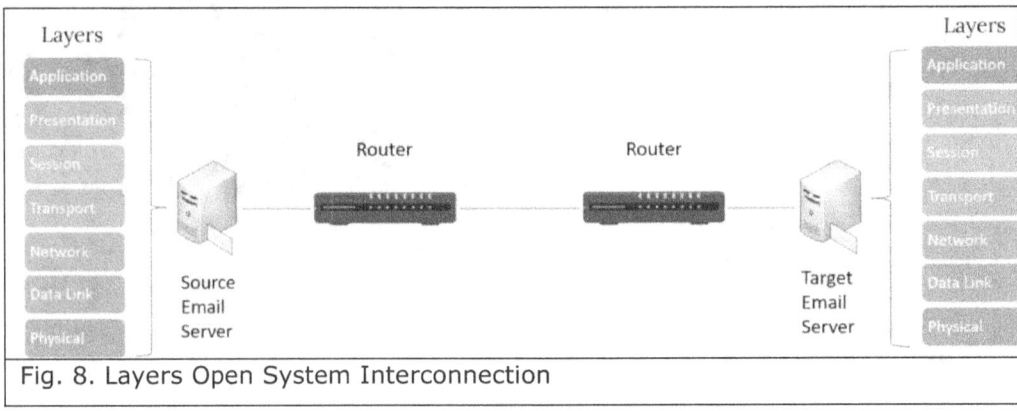

Fig. 8. Layers Open System Interconnection

The layers according to the Open Systems Interconnection-model have the following purpose:

layer	purpose
application	the email client uses SMTP protocol to communicate with the email server

---

[25] https://www.baeldung.com/cs/tcp-max-packet-size

presentation	converts the mail into ASCII and images
session	establishes and maintains the connection with the target server
transport	splits the message into multiple packets and adds port information of both source and destination servers.
network	defines the routing path of packets by adding corresponding IP addresses
datalink	prepares the packet to transfer over the Ethernet
physical	transmits the frame over the physical connection like LAN-cable, WiFi or broadband
https://www.baeldung.com/cs/tcp-max-packet-size	
Tab. 5 Layers Open System Interconnection	

"When the packets arrive at the Physical Layer of the target network, each layer at the target email server processes the packets to retrieve the data and show the email in the target Inbox. An e-mail is divided into several packets with each a header and a payload. The packet contains information for directing them to the target address and information for checking transmission error and data integrity. The header has the port & IP-address of the source and target. A frame is a combination of packets where each packet can take a different route. The maximum size of a packet is 64 Kb restricted by the maximum transmission unit. The larger is the packet size, the more probability of packet loss".[26]

Transport like 3D Scanning and Printing

Mail is hard to compare to a body. I do see another option. Instead of moving an object, you could only transfer the information on the way the object is made along a data line and assemble it on another location from raw material. For example a 3D printer works this way. This can work for materials like metal, stone, plastic or even organic pasta. We can relate this to 3D scanning. The information to make a 3D print could come from a 3D scan which is recorded as 3D parameters and/or right away from the 3D parameters. We can call this the product information. There is a relation with computer aided design and computer aided manufacturing (CADCAM). CAD and a 3D scan create the 3D parameters. CAM uses the product information to create or 'print' the

---

[26] https://www.baeldung.com/cs/tcp-max-packet-size

product remotely from the scan location. It is also possible to make a metal object from a block of metal like an aluminum wheel. Instead of assembling a product in various steps from many different components, all you need is the basic material to 'print' it. It bridges the gap that existed between designing a product by hand or on a computer and instructing a machine to produce the item. No need to swap disks with information anymore as the computer and the machine are in this way connected.

3D scanning analyses an object or a body and creates three-dimensional data consisting of multiple point on the surface of the object to construct a 3D model. Even colors can be registered[27]. Obviously the scanner has a limited view of the body and either the scanner must rotate, multiple scanners must be used as cameras in an array or the object itself must rotate in order to get a full image. Then these different scans have to be combined or aligned for the 3D, computer aided design or digital surface model. A variety of technologies exists like radiation or light methods and contact versus contactless techniques. A principle is: **scanning uses a contactless technique with safe radiation.**

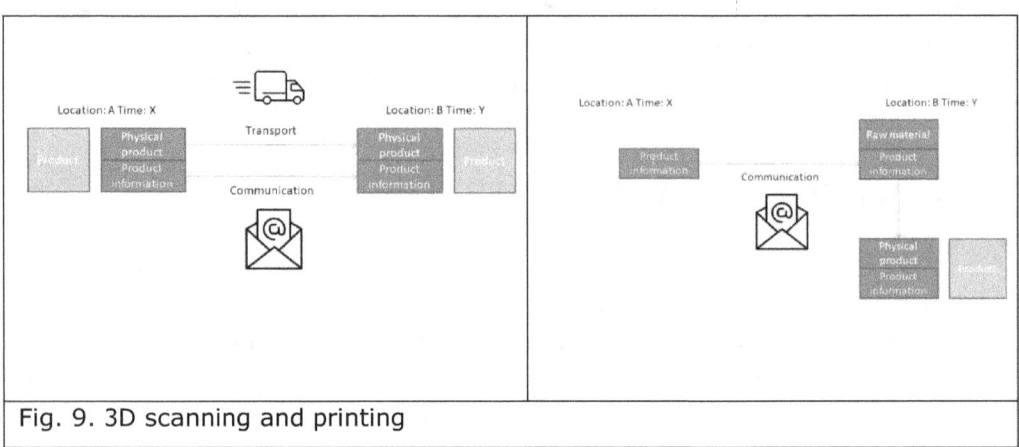

Fig. 9. 3D scanning and printing

We can philosophize on the materials that a human body is made of to see how the body could be reconstructed with the information at another location.

---

[27] https://en.wikipedia.org/wiki/3D_scanning

About 99% of the mass of the human body is made up of six elements: oxygen, carbon, hydrogen, nitrogen, calcium, and phosphorus. Only about 0.85% is composed of another five elements: potassium, sulfur, sodium, chlorine, and magnesium. All 11 are necessary for life. The remaining elements are trace elements, of which more than a dozen are thought on the basis of good evidence to be necessary for life.[28]	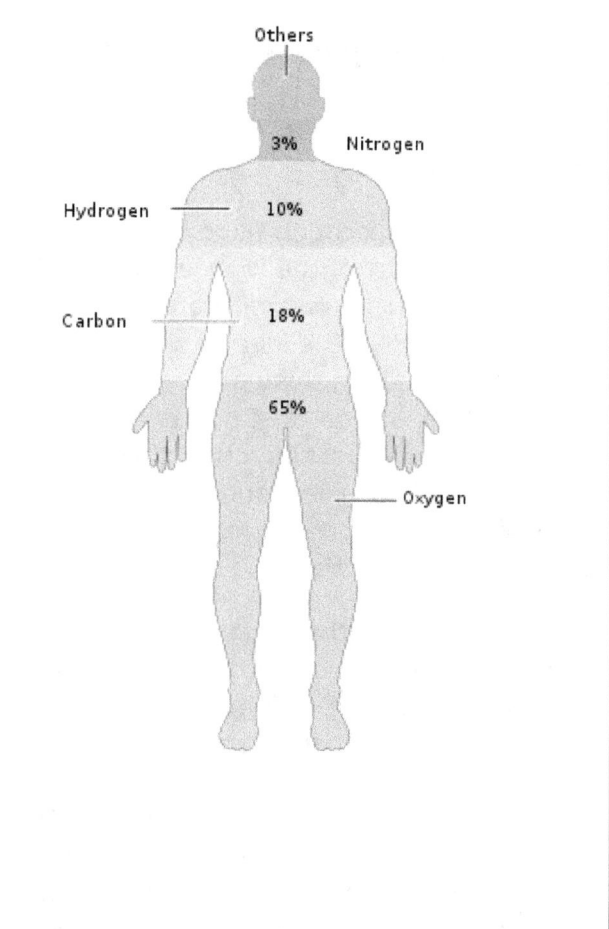
Deoxyribonucleic acid (DNA) is a polymer composed of two polynucleotide chains that coil around each other to form a double helix. The polymer carries genetic instructions for the development, functioning, growth and reproduction of all known organisms.[29]	

Fig. 10. Composition of the human body.
https://en.wikipedia.org/wiki/Composition_of_the_human_body#Molecules

In a very simple and probably not very accurate way we can say that a body could be reconstructed at another location if we know the materials it is made of (body cells) and we know how these materials are related by the DNA-profile. A principle is: **the DNA-profile is one of the data used to copy and send the materials of the body**.

---

[28] https://en.wikipedia.org/wiki/Composition_of_the_human_body#Molecules
[29] https://en.wikipedia.org/wiki/DNA

4. Travel in Time and Location

4.1 *Spacetime*

So far time and location have been considered as separate dimensions where in the part of the 'Concept of time' the time zone on earth has been mentioned as part of the absolute denotation on time. In the part on 'Time machine' the idea of space as a fourth dimension of time has been introduced. In this paragraph I will go more into detail on this relation. I have tried to stay away from the complexity of the theory of relativity and introducing other theories that take away from the subject of this book. I cannot though completely ignore the impact of this theory on time and location so a few words are appropriate without a complete explanation of the theory.

Spacetime is a mathematical model which combines the three dimensions of location with the dimension of time like stated before in the 'concept of time'[30]. Time can be seen in this model as the measurement of when events occurs in the universe. This definition is a little different than time used in this book as I do not talk about events but about being somewhere sometime. Neither do I speak about the universe which can be seen as a subset of space like it has been described before. Spacetime deals with how **observers** perceive **where** and **when** events occur. I can relate these three concepts as follows:

- Observer: the person ('traveller') who is involved and travels through time and/or location
- Where: the dimension of location
- When: the dimension of time

If we look at the two options of time travel being the time portal and the time machine (the last includes the Delorean), there are some remarks to be made regarding the observer. With a time portal there is not much to observe so this option does not apply during the travel but only at the event or situation before and the situation after. With a time machine equipped with a 'window' to the outdoors, there might also an event or period of events during the travel.

---

[30] https://en.wikipedia.org/wiki/Spacetime

Minkowski introduced the special relativity in 1908 in this model of the four dimensions. I will not go into the general theory of relativity which deals with mass and energy as this is not relevant to the context of this book. Special relativity states: *"the observed rate at which time passes for an object depends on the object's velocity relative to the observer."* (same footnote below). Therefor time is seen as dependent upon location. Time dilation refers to the fact that time passes at different rates for different observers, depending on their relative motion or positions in a gravitational field.[31] I do not see a real impact on the subject of this book as experiencing the past or the future.

4.2 Comparison of means

We can compare the means in the previous chapters in this way:

means	time	location	science
Time Portal	years and centuries	optional	fantasy
Time Machine	years and centuries	no change	fantasy
DeLorean Back to the Future	years and centuries	no change	fantasy
Star Trek Transporter	almost immediately, see mail	anywhere	fantasy
Cocoon or Pod	years and centuries, only forward	you do not move but your surroundings move	promising
e-mail	almost immediately; electromagnetic waves have a speed of about 90% of the speed of light (300,000 m/s[32]) or 270,000 m/s	anywhere	proven
3D Scanning and Printing	almost immediately, see mail	anywhere	proven
Tab. 6 Means of travel in time or location			

4.3 Summary of the principles for time and location travel

We have derived the following principles that apply for travel in time and/or location:

---

31 https://www.livescience.com/what-is-time-dilation
32 Or to be more precise the speed of light is 299,792,45 m/s in vacuum.

	principle	dimension
1	You must use a standard notation for time and day including the time zone.	time
2	You cannot change or create an event in the past or in the future.	time
3	You cannot remember details of the future or the past at the moment you were there.	time
4	When travelling in a cocoon for a long period, the person must be monitored for health and be fed (orally) before and or during the trip.	time
5	The location of destination must not be occupied by a subject and also needs a receiving station.	location
6	Scanning uses a contactless technique with safe radiation.	location
7	The DNA-profile is one of the data used to 'copy and send' the materials of the body.	location

4.4 Travel to Mars as an example

Looking at what we have discovered in the previous chapters on time and location I finish with a possible trip to Mars as the most likely way to travel in time and location. I will include each of the relevant principles that have been derived before in this chapter.

Let's look as an example what it would take to travel to Mars in a cocoon within a space ship. As Mars looks a little like the earth and is far away from us, it is not unthinkable that mankind wants to go to Mars and might reach it someday. A practical aspect is the distance from the earth to Mars which is variable as both planets move in a non-circular orbit around the sun. It is therefore very important to pick the right time to start such a journey. A regular timetable like in the movie 'Total recall' is there for very unlikable (principle 1). Only every 26 months the planets are in a good position to travel in an energy efficient way. The distance can vary between 55 million kilometers to 400 million kilometers! At the speed of light (300,000 meters an hour) this would take between three hours and 22 minutes. The trip to Mars of the Perseverance Rover in 2020/2021 took about 7 months at an average speed of 39,000 kilometers an hour. The Apollo 10 had a speed of 39,897 km/h so there does not seem to be a lot of progress in speed in these decades.

Another question is whether men would travel to Mars and back or just one way to set up a colony in Mars. That would reduce the time involved and the risk in a severe way. The persons need to be monitored automatically (principle 4). We assume the universe is empty but there is dust and there are rocks and planets that must be avoided by the spaceship. The speed of the spaceship will push any air forward and create 'room' for itself (principle 5).

The cocoon could be a soft shell as described before when there is no artificial gravity or a hard shell if there is artificial gravity in my opinion. Another issue that is interesting, though it has nothing to do with time or location, is the reuse of energy and material as this is something that has a limit to take from the earth at the start.